CONQUER THE INTERNET

CONQUER THE INTERNET

By IAN LEWIS

Illustrated by
Tim Benton

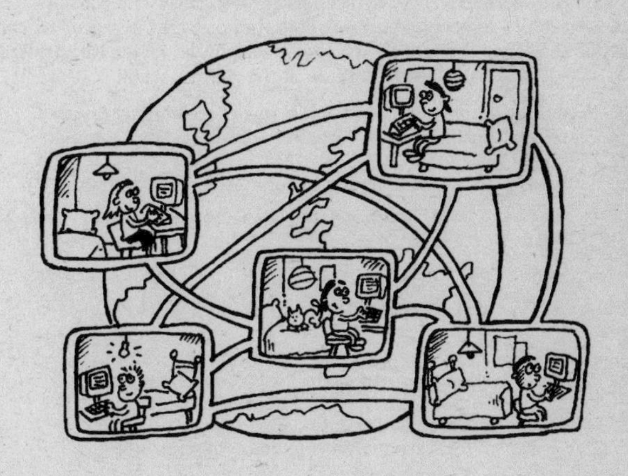

OXFORD
UNIVERSITY PRESS

OXFORD
UNIVERSITY PRESS

Great Clarendon Street, Oxford OX2 6DP

Oxford University Press is a department of the University of Oxford.
It furthers the University's objective of excellence in research, scholarship,
and education by publishing worldwide in

Oxford New York

Auckland Bangkok Buenos Aires Cape Town Chennai
Dar es Salaam Delhi Hong Kong Istanbul Karachi Kolkata
Kuala Lumpur Madrid Melbourne Mexico City Mumbai Nairobi
São Paulo Shanghai Taipei Tokyo Toronto

Oxford is a registered trade mark of Oxford University Press
in the UK and in certain other countries

Series devised by Hazel Richardson

First published in 2001

British Library Cataloguing in Publication Data available

ISBN 0-19-910768-8

3 5 7 9 10 8 6 4 2

Printed in Great Britain by
Cox & Wyman Ltd, Reading, Berkshire

Contents

BEFORE YOU START

Everyone has heard about the Internet by now. Except maybe for people who have spent the last ten years living under a table in the dark in a cottage on a deserted island with their cat.

But, although everyone – just about – has heard of the Internet, a lot of people don't actually know quite what it is. Or what they can do with it.

What's it *for*? The simple answer is that the Internet is for whatever you want. Some people say it's like an enormous library. That's sort of true; but it's a library that looks more like your bedroom than the library in town. It doesn't just have books in it and you know that things are there somewhere, but it's not always that easy to find out quite where.

Really, the Internet is more like the telephone. It's just a way of connecting all sorts of computers together, just as the telephone is a way of connecting all kinds of people together. What you do with the connection is up to you.

In this book, you'll find out about:

- what the Internet is, and where it came from
- how to get the best out of your connection
- the most important bits that make up the Internet (there's more to it than the World Wide Web)

- lots of brilliant stuff you can do, and how to find it
- how to become part of the Internet – building your own web site, and the things you can do with it

Don't forget...

With all the goodies on the Internet, it's easy to get carried away and spend hours looking at stuff. If you have a telephone line for the Internet only, and your household pays a fixed amount every month for any time they spend on the Internet, that's fine. But many people still have to pay for telephone calls to the Internet, and this can be costly.

You also need to bear in mind that you could be holding up the telephone line while your sister or brother is trying to call to get themselves a lift home, or your Great Aunt Aggie is trying to tell you that you've inherited £1 million.

CHAPTER ONE

WHAT IS THE INTERNET?

During the 1960s, fear of a possible war between Russia and the USA meant computers became more and more important, and so did the communications between them. Military leaders were worried that if a telephone line were bombed, messages between computers wouldn't get through. The solution was the Internet. The Internet connected most of the big computers in the USA by all possible routes, so that any message could get through, simply by taking a different route if it had to.

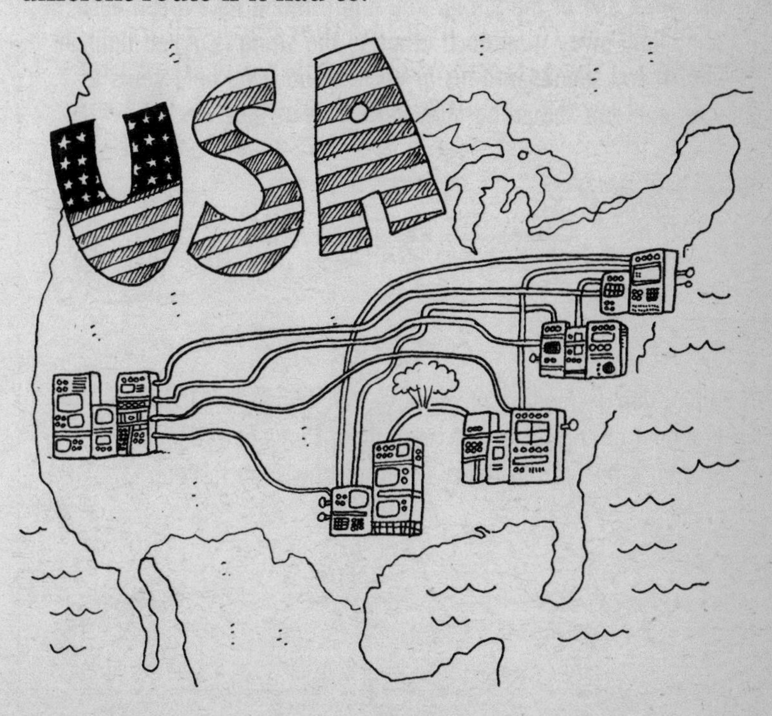

Be an Internet expert:
MAKE YOUR OWN INTERNET

Try making your own telephone network to prove how the Internet works.

WHAT YOU'LL NEED

 three friends

 eight empty tin cans

 four bits of string

WHAT TO DO

First of all make yourself four sets of tin-can telephones. Make a small hole in the bottom of each can, push the string through it, and knot the string so it can't slip out again. Do the same at the other end of the string. You and a friend take a can each and stand away from each other so the string is pulled tight. If your friend speaks into his or her can, and you hold yours to your ear, you should be able to hear what they say.

When you've made four of these telephones, set them up between all four of you in something that's like a square – you to Sam, Sam to Azim, Azim to Amy, Amy to you.

➤

Now, you want to talk to Amy. But you can't. (Pretend a missile has blown up the line between you.) But never fear, because you've set up your telephones like the Internet. You can pass your message through Sam and Azim to get to Amy.

For a long time, this early Internet was only used by the military and by universities. However, that all changed in 1992, when Tim Berners Lee came on the scene. Lee was an Englishman working at CERN (a nuclear research organization) in Switzerland. He suggested a way of using something called 'hypertext' to make scientific documents easier to read. This was to be the beginning of the World Wide Web, and the Internet as we know it.

What's hypertext? ....

Hypertext is a means of linking things together in a way that means you don't have to start at the beginning, go on to the end, and then stop. Here's a hypertext for Tim Berners-Lee.

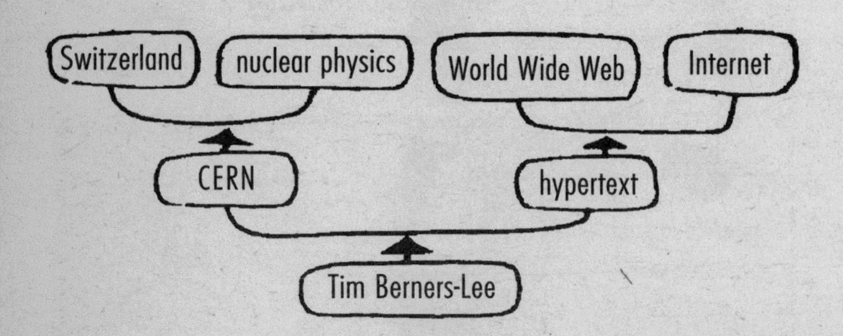

For instance, this paragraph about hypertext is separated from the rest of the book's text. If this book were hypertext, you wouldn't have been able to see what you're reading now until you pressed your computer mouse button on the underlined word above ('hypertext'). Then you would have gone straight to it. If it's used properly, hypertext is a very clever way of letting you read what you're interested in without making you slog through lots of stuff you don't care about.

How it works

But how does the Internet get us to this stuff in the first place? Think about this. If you have two (or more) computers in your house, it's easy to link them up so that they can both use the same printer and the same modem, and so on. You can create a network. When thousands of computer networks around the world are linked, you've got the Internet. Whenever you connect, you, too, are part of the Internet.

The Internet is not just one thing, it's made up of millions of different pages, sounds and images from millions of different people. Some are put there by big companies; others are from people like you who just put up the things that interest them for others to see. Some are brilliant, wonderful and useful. Some of them are definitely rather strange.

This book is about finding your way around in this chaos, and getting the most out of it.

CHAPTER TWO

THE NEED FOR SPEED

In the early days of the Internet, most of the stuff on it was basically just text, which doesn't take very long to send down a telephone line. Things have changed since then. More and more web sites are being built as multimedia experiences – with sounds, music and moving pictures.

When you want to transfer something from the web site you're looking at, that data has to travel across the Internet, down your telephone line, and into your computer, where it is reassembled (from a stream of 0s and 1s) into the words, pictures, music or video you have requested. Text moves quickly. Quite a large text file might still be only 2Kb (two bits of data per second) or so. Pictures are bigger. Even a small picture can easily be 30Kb. Music files are around half a million times bigger than a 2Kb text file. And video takes even more room.

This is why having a fast connection is becoming more and more important to Internet users.

Be an Internet expert:
MAKE A FAST CONNECTION

Try this experiment to show how different connections affect the speed at which things move.

WHAT YOU'LL NEED

 sticky tape

 an ordinary drinking straw

 piece of garden hose

 the extension tube from your vacuum cleaner (or even a piece of drainpipe if you can find one)

 a funnel

 three or four different sized containers: a little glass, a jam jar, a milk bottle ➤

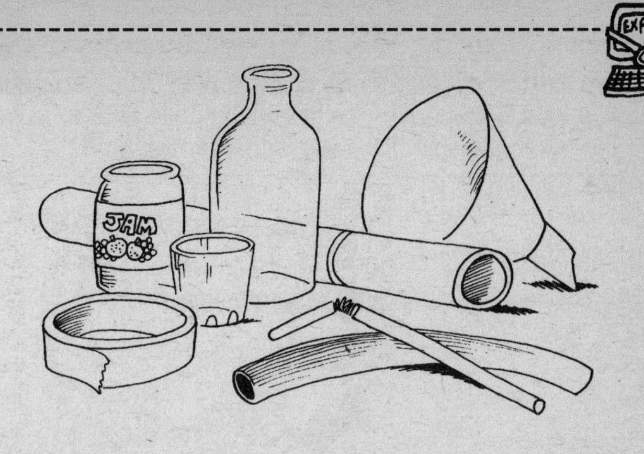

WHAT TO DO

Fill all the containers with water. Each one represents a different type of file. The little one is the text file, the jam jar is the graphic, and, you've guessed it, the milk bottle is the video file. The water in the containers is the data you want. Now you use the various pipes to represent different speed connections to your computer.

Connect the funnel to the drinking straw with sticky tape. Then start pouring. Try pouring the water out of the containers and into the funnel at the same rate. You'll probably find that you can just about manage to keep the text file in the little glass going through, even if it's ever so slow. But if you try tipping the water out of the bigger containers at the same rate it all goes everywhere.

When you attach the garden hose to the funnel, it should take the text file and the graphics file quite well, but I bet you spill the water from the milk bottle (video file).

With the drainpipe attached to the funnel, you should be able to just pour everything in with room to spare.

WHAT YOU PROVED

The drinking straw is like an ordinary telephone line with an ordinary modem. If you have an ISDN line, like BT's Home Highway, you have a connection more like a hose-pipe. But you can't handle video until you get a nice fat 'drainpipe', such as ASDL, bringing data into your home!

Revving up!

Three forms of 'drainpipes' (ways of speeding up information flow) are already available to some lucky people. They are slowly spreading, so mere mortals like us should be able to use them one day soon.

Satellite One of the fastest ways of getting the Internet is through part of a digital satellite TV channel. Basically you put a satellite receiver card in your computer, connect it to your satellite dish, and off you go.

Of course, you can't talk back to the satellite, so everything you send to the Internet – like when you click on a link – has to go through your telephone line. That means you still pay telephone bills (or someone does) and the connection is only fast one-way.

ADSL This system is better. ADSL stands for Asynchronous Digital Subscriber Line. It's a very clever technology that makes an ordinary telephone line very fast indeed by sending the data in a radio signal along the wire. In theory it can be as fast as the satellite link.

Even more important: the connection is always on. (No, you don't have to be sitting at your computer all the time.)

Cable modems Cable TV companies are beginning to offer incredibly fast Internet access through their TV cables, too.

Home delivery

So now you know how data gets to you, but how does the Internet know which data to send to you, and which bits to send to Pascale in Paris, or Micky in Munich? The answer is that the data travels in packets.

Every time a request from you goes out across the Internet, or a piece of a web site comes back, the data is cut into little bits. Each bit is labelled with something called the IP address of your computer (that's those numbers you've probably seen when the Internet is being particularly unfriendly – something like 192.168.0.2). That way the Internet knows where to send everything, and your computer – or the other one you're talking to – puts all these bits back together again in the right order as they arrive.

That's why, no matter how fast a connection you have into your home, the Internet can still be slow at times. Your packets are jumbled up with everyone else's, and you've just got to wait for them to arrive.

Be an Internet expert:
SHOW HOW INFORMATION MOVES

Try this experiment to show how Internet hold-ups occur.

WHAT YOU'LL NEED

 coloured sweets

 the tube from the sweets (or a rolled-up piece of paper)

WHAT TO DO

Divide all the sweets up into their separate colours, and decide which colour you want to be. The other colours represent information going to other people at their computers.

Now pick up the sweets one at a time, but mixing the colours, and put them through the tube. The sweets are data leaving different people's computers and going across the Internet.

They all leave your computer (the sweetie tube) in a sensible order, but the colours all get mixed up. You can still tell which sweets belong to each person because of the colour, but they need to be sorted out again.

When the Internet seems slow, what's happening is that you have to wait for other people to get their sweets before you get the next few of yours.

Tips for getting top speeds ....

Here are some things you can do to improve the speed of your Internet connection.

- Avoid rush hours. When the Americans wake up, at around lunchtime in the UK, you'll find that the whole Internet slows down.

- If you're using a modem (and not something which gives a faster connection like ISDN or Cable) for your Internet connection, make sure that you have a modem that can run at 56Kb per second (56,000 bits of data per second). Any money you might have to spend to upgrade your modem, you'll save in shorter telephone calls.

- Make sure your telephone line is as clean as it should be. If you hear crackling when you're talking on the phone, then your Internet connection will probably be slower, too. Your telephone company can sort this out. While they're at it, tell them that you're using the line for the Internet and make sure that they're not splitting it with someone else at the exchange.

- A faster computer can make a difference. Web pages are actually just a set of instructions for how the page is supposed to be displayed. A faster computer does the work more quickly.

- Don't just settle for any old Internet Service Provider (ISP). Some are faster than others. And some are faster at different times of the day. For example, one that specializes in business connections is likely to be slower during working hours.

CHAPTER THREE

TAKE THE INTERNET APART

You probably already know about the World Wide Web. In the last few years www.dot.com addresses have sprung up all over the place like fleas on a dog's back.

Some people only ever think of the World Wide Web when they talk of the Internet, but there's more to the Internet than just that.

E-mail

E-mail (electronic mail) is so-called because it works in a similar way to the postal system – and come to think of it, you get just as much junk-mail from the Internet, but that's another story.

Actually, in some ways e-mail is not quite as efficient as the mail that comes through your door. If you send someone an e-mail, it doesn't go straight to the person you've addressed it to, but to their local 'post office' – normally their Internet Service Provider's computer. And it sits there until they collect it. If you send a letter by normal mail, you can be pretty sure they'll get it. But you can't be so sure about an e-mail, because the person you have written to has to make the effort to collect it.

HOW TO DO E-MAIL

Basically there are two ways of dealing with e-mail. You can do it through an 'on-line' web-mail service, like Hotmail or Yahoo. Or you can do it using a mail program on your computer, where you only connect to the Internet to send or collect your mail. Everything else you do 'off-line' (while you're not connected to the Internet).

The good thing about on-line services is that they work through your web browser. This means you can sign up and manage your mailbox easily, and you can read and send mail from any computer in the world.

On the other hand, being 'on-line' means you are connected to the Internet while you do everything. Telephone bills can mount up quickly – and, even if you pay a fixed rate for Internet calls, you're still taking up the telephone line needlessly.

Be an Internet expert: SEND AN E-MAIL

WHAT TO DO

Connect to the Internet and go to msn.co.uk where you can sign up for a Microsoft Hotmail account. You don't have to use it again, but it's a good way to see how web-based e-mail works.

➤

It's best to think of a name before you sign up, because your real name is probably already taken.

Click on the line 'sign up for free e-mail' and follow the prompts. When you're signed up, send yourself an e-mail, either to your new Hotmail account or to the e-mail address you have from your Internet Service Provider.

Web-based e-mail is relatively new. All e-mail used to be handled off-line, and most of it still is. The good thing about off-line mail is that you're not paying telephone charges while you read and write it. But you do need to use a special e-mail program.

E-mail program choices

These are some of the best e-mail programs, and the great thing about them is they are all free!

A lot of people use *Outlook Express*, which comes as part of Microsoft's Internet Explorer.

➤

Eudora (www.eudora.com) is also very popular, has been around for a long time and there are versions for PC and Macintosh.

Pegasus (www.pegasus.usa.com) is not only easy to use, but has some very advanced features if you want to make use of them.

(All of these programs crop up regularly as free disks on the covers of computer and Internet magazines.)

WHAT CAN YOU DO WITH E-MAIL?

You can do all sorts of things with e-mail. It's a painless way of keeping in touch – writing letters no longer has to be the kind of torture you dread after Christmas or birthdays.

E-mail is easy and quick. You send it, and (usually) it's there straight away. If you're lucky, you get a reply in minutes (but don't rely on it). So you write in quite a different way than you would on paper. E-mail is much more casual and chatty – you can just say stuff.

You can also send things like pictures or sounds to your friends. Most e-mail programs have a button labelled 'attach' or something like it. This attaches something, like a picture or a sound, to an e-mail message.

Be an Internet expert: SEND A PICTURE BY E-MAIL

WHAT TO DO

Write an e-mail message to yourself. Before you send it, find a small picture somewhere on your computer – like a piece of clip art. Click the 'Attach' button on your e-mail program, and add the picture you've found. Then send the mail. With any luck the message you receive will say it has an attachment. Save the attachment, and you should be able to see your picture again.

Why do you need to 'attach' stuff like pictures or sounds? Well, all e-mail is just text. All letters – nothing else. So you can't just dump pictures, sounds or music into an e-mail, you have to 'attach' them to it. It's a bit like sending a letter with a parcel attached to it.

When you attach something to an e-mail, your mail program encodes the picture or sound. This means it turns it into something that is just plain text. Now it can travel with the main text until it reaches its destination computer where it is turned back into sounds or pictures.

E-mail errors!

There are some things you should know about adding attachments to e-mails. Firstly, it doesn't always work. There are two or three different ways of attaching files to e-mails, and not all mailers understand all of them. The best way of making sure it will work is to check that the person you're sending the attachment to uses the same e-mail program as you.

Secondly, if you use Windows, and you're sending something like a screen saver to a friend, it will only work in Windows. If your friend has something else, then you're wasting your time.

Thirdly, be considerate. Big attachments can take a long time to download. It's very rude to send someone a big picture (or anything), no matter how stupendous it is, without asking first.

Lastly, be very careful of viruses. Never open an attachment you're not expecting – at least without virus-checking it first with a virus-checking program like the McAfee or Norton Anti-Virus programs. This goes for Microsoft Office documents, too, because they can contain macros (little programs buried inside documents) that can do horrible things to your computer.

MAILING LIST MAGIC

You can get lots of really interesting stuff sent straight to you by e-mail. Many web sites have 'Join our mailing list' buttons. You just fill in your e-mail address and press the button, and stuff like daily news headlines or e-mail magazines (known as an 'e-zines') comes straight to your computer.

Mailing lists are available from http://www.neosoft.com/internet/paml/ and http://www.liszt.com

Be an Internet expert: CREATE AN E-MAILING LIST

You're going to send a group of friends the same message, all at once.

WHAT TO DO

First write your message. Then put your friends' e-mail addresses in the 'To:' box on the e-mail program. Put a comma between each address. Then press 'Send', to send the message to everybody at once.

Put yourself on the list too, so you can check it works. Oh, and make sure that the people who get your message know how to tell you if they don't want to be on your e-mailing list!

ANOTHER WAY TO DO IT

There's an even easier way of doing this. Create a mailing list (or 'group') and put all the addresses in the list. Then you only have to put the list name in the message to send it to everyone at once. Look in your e-mail program's 'Help' menu to tell you exactly how to create a list.

NETIQUETTE – E-MAIL DOS AND DON'TS

Some people make a lot of fuss about so-called netiquette – net etiquette – though you read a lot less about it now than you used to. But really there's only one thing to say about how you should behave on the Internet – just act like a human being!

Because you type e-mails, or 'talk' to someone in a chat room by typing, it often doesn't seem quite real. You're kind of insulated from the other person. This means that some people say things they'd never dare to say usually.

A golden netiquette rule is never write anything that you wouldn't say to someone if they were in the room with you. Also, if someone makes you cross, don't write straight back – wait until you've cooled off. If you don't, things might get out of hand!

And, however excited you might be about your new web site, or the list of jokes you've just found, or whatever, never commit 'spam'. Spamming is sending a message to thousands of people at once. You'll get some spam practically as soon as you're on-line, and these messages are not only incredibly annoying, they also clog up the Internet. Most Internet Service Providers forbid them, and if yours find that you've been doing it, they'll close your account (they say so in the small print you didn't read when you signed up).

Oh, and one last thing. DON'T WRITE EVERYTHING IN CAPITAL LETTERS, BECAUSE IT MAKES IT LOOK AS IF YOU'RE SHOUTING.

Internet addresses

Addresses on the Internet (like you@penguins.co.uk) are a lot more user-friendly than the numbers they represent (remember 192.168.0.2?), but they can still seem a bit confusing. It's all quite logical really. The bit before the @ is anything you want to be called. The bit immediately afterwards is the *domain* name (which might be your ISP) and the rest is the type of domain it is.

Sometimes you can tell useful things from an e-mail address. For example, the domain type *co* means a company; *org* means an organization, and *gov* means, yes, you've guessed it, government.

The domain '.com' was originally meant to be for commercial (business) organizations in the USA. Since then, having a '.com' domain has become cool, and all sorts of people have one.

The trouble is that the Internet has grown at such a rate that the original system can't cope any more. So you'll see all sorts of other funny names, now.

The best thing is just to be careful with your typing.

FTP

FTP stands for 'File Transfer Protocol', and it's a way of copying stuff that isn't text – like programs and music – across the Internet. Web browsers can do it for you, but it's usually quicker to use a special ftp program to download stuff. You can get them free from magazines, or download them from the Internet. Most ftp programs look a bit like Windows Explorer, and you can use them in much the same way, just copying things from the Internet to your computer. (But remember to check for viruses before you run it or install anything.)

Be an Internet expert: DO AN FTP TRANSFER

Many ftp sites are open to the public. They'll let you log on 'anonymously' using just your e-mail address.

WHAT TO DO

Every ftp program has Microsoft's address already in it. Connect to the Internet and go to ftp.microsoft.com. You'll see a list of folders just like the ones on your computer and a file called something like 'index.txt' or 'readme.txt'.

Click on one of these to read or copy it to your computer. Now it will tell you how to find other stuff. Congratulations! You've just done your first ftp file transfer. Now you can go back and download a program you really want.

NEWS

The newsgroups on the Internet have nothing to do with the news you see on TV. Newsgroups are simply discussion groups.

There are newsgroups on practically every subject you can imagine, and almost certainly a few you can't.

Be an Internet expert: JOIN A NEWSGROUP

WHAT YOU'LL NEED

 a news reader (a specialized kind of e-mail program. Outlook Express is a very good news reader.)

WHAT TO DO

Connect to your news reader at your ISP. The first time you connect, it will download a list of up to 65,000 news groups. Choose one and connect to it.

To save your telephone bill, you can get your news reader to download the message subjects first. Then you simply decide which messages you want to read before going back to get them.

IRC

IRC stands for Internet Relay Chat. It's a way of chatting to people across the Internet. It's a brilliant way of making friends all over the world. There are various ways of doing it. Most of them are 'text based' – that is, you type words in – but you can actually talk to people with real words if you have a microphone on your computer, and use a chat program such as Netmeeting (www.microsoft.com/windows/netmeeting), Firetalk (www.firetalk.com), or mIRC (www.mirc.co.uk). All chat programs have lists of chat groups you can visit, to 'talk' to people if you want to, or just to listen.

Chat can be fun because nobody knows who you are. But you've got to be careful, because you don't know for sure who anyone else is either.

Be an Internet expert:
JOIN A CHAT ROOM

Have a go at chatting straight away – or at least see what it looks like.

WHAT TO DO

Click on 'Search' on your web browser.

Enter 'Chat' into the search box and press the search button. You'll get a page listing some places you can go to chat.

Choose one that looks interesting and go there. You have to choose a nickname for yourself before you join, so think of one first. And there you are. You don't have to say anything. You can just read what others are saying for a while. Join in when you're ready.

What not to do in a chat room

- Don't reveal your address, telephone number, school name, send a picture or arrange to meet anyone you know through the Internet.
- If someone says or writes something in a chat room or e-mail that makes you uncomfortable, report it.
- Keep your password secret or people can pretend to be you.
- Never give credit card or bank details without first checking with your parent or carer.
- Never respond to nasty messages.

Remember, people don't always tell the truth on-line.

Instant Messaging

Instant Messaging is a bit like chat. The idea is that you have it on all the time you're connected to the Internet. So, if one of your friends is connected and feels like talking to you, they can get in touch instantly. ICQ, AOL and Microsoft all offer ways of doing this.

CHAPTER FOUR

FINDING THINGS

This is probably the most important part of the book. Actually it's the most important part of your life, because if you know how to find things out for yourself, then you can do anything you like – well, pretty much anything.

Search engines

Search engines are web sites that help you find things on the Internet. Some of the biggest ones are Yahoo, Altavista, Google, Lycos and Excite. Search engines work in one of two ways: like indexes, or like catalogues.

Most work like indexes. The search engine sends sorts of electronic robots called 'spiders' around the Internet, looking at all the words of all the pages they find and compiling them into one giant index.

Yahoo is quite different. When someone asks for their web site to be listed in Yahoo, it's visited and checked out by a real person; and then put in the appropriate category. So Yahoo is more like a giant catalogue.

HOW TO USE SEARCH ENGINES

You do the same thing with any search engine. Go to their page on the web and you'll see a box on the screen. In the box, you type a few words to do with the subject you're looking for. That's it. After a few seconds you get back a list of pages that the search engine thinks are to do with the words you entered.

Some people moan that all they ever get off the Internet is rubbish.

There'll always be some rubbish, but if you don't get good stuff as well, you're not using the search engines properly. All you need to do is to think harder about the words you are asking for.

Be an Internet expert: KNOW HOW TO FIND THINGS

Try a few of these things out. Let's say you're looking for stuff about dates – the kind that grows on trees. It's no good just typing 'dates' into the search box. You'll get all sorts of stuff to do with calendars, and some on finding people to go out with you! You have to be more specific.

Think of some words that are only likely to appear in pages about date as a fruit. 'Fruit' might be a good one. So might 'tree'. You could try 'camel', but you might get some strange results!

By using 'fruit' and 'tree', you should find a couple of references that lead you to trying 'date' and 'palm'. (Dates grow on palm trees.) This should get you a page full of exactly the right stuff.

Search engine tips

Here are some tips to help you improve your Internet search skills.

Don't use capital letters.

Enclose words that must be together in quotes. This can be particularly helpful with names. If you're looking for *s club 7* – and a lot of people do – the band will probably come quite high up your list of results anyway. But if you put the name in quote marks like this – *"s club 7"* – you'll be sure of getting even more.

Use the minus sign in front of any word you want left out of the search. For example, if you were looking for pages about geology (the study of rocks) and you did a search for 'rock', you'd find thousands of music entries and not many about stones. If you search for 'rock –music', the geological pages should come at the top of the list.

Meta-search engines

Meta-search engines – funny name, but very useful things. None of the search engines knows about everything on the web. So, to be sure of finding all the best stuff, you really need to use more than one search engine. But how?

Meta-search programs ask several different search engines to search at once. It's a bit like looking in several different encyclopedias simultaneously.

Meta search engines usually run on your computer. There are several of them, and the best way to choose one is to try them all out to find the one you like best. Good free ones are Copernic (www.copernic.com), and Web Ferret (www.ferretsoft.com).

Grabbing sites

One last dead-useful general tool that saves money and time is a 'site grabber'.

Sometimes you find you visit a really useful page and two days later, you decide you want to look at it again, so you have to re-connect to the Internet and go back to the same place (and pay more telephone charges). Well, there's a way round this. You can use the newest versions of web browsers like a kind of video recorder for the web. When you look at a web page, it's stored temporarily on your computer.

Be an Internet expert: STORE A WEB PAGE

First, use your web browser to store a web page.
Find the 'off-line' option (normally hidden on the 'File' menu), and choose to work off-line. Now instead of going on to the Internet to get the page you want, the browser will look for it on your disk instead.

Type in an address or choose something from your 'Favourites' menu that you've recently visited. You should see the page appear without having to connect to the Internet.

Downloading sites

You can download whole web sites to your computer, and look at them whenever you like with programs like Webwacker (www.bluesquirrel.com). This can be really useful if, say, you've found a huge site about penguins, and you want to read all of it without having to pay massive telephone charges, or hogging the telephone line. That's if you're really into penguins, of course.

These helpful programs all work in much the same way.

1 Put in the address of the web site you want to copy to your computer. Decide how much of the web site you want to copy. You may want to copy all of it or just a particular section – maybe just the bit about penguins from a general site about animals, for example.

2 Tell the program how many levels to grab, going on from the page you start at. The page itself is level 1. If you took everything that it linked to, that would be 2 levels. If you went on another stage, 3 levels, and so on.

3 Next, decide whether to let the program collect from pages on other sites that this one might link to. Normally the best answer here is no. If you're not careful, you could end up downloading the whole Internet to your computer.

4 You can often tell the program how long to connect for, and even when to connect, so that you can come down in the morning and find the morning paper all ready to read on your computer.

CHAPTER FIVE

COPYRIGHT

This is a very short chapter. A lot of people seem to think that everything on the Internet is free and they can do what they like with it. This isn't true.

Imagine you've spent all weekend working on a really brilliant design for a T-shirt. A couple of days later you're out shopping and see dozens of people wearing your T-shirt design. Someone just took it without asking. You wouldn't be very happy, would you?

It doesn't matter whether the person who stole the design made money from it or not – although that certainly makes it worse. The point is that they stole something that belonged to you.

Any kind of work, including web pages and music and software belongs to someone. Some people or organizations say you can use stuff on their site for free. But it still doesn't belong to you. So think before you just take stuff. Unless there's a big sign up saying you can have it, ask first. OK?

CHAPTER SIX

THINGS TO DO ON THE WEB

This whole chapter is meant to give you some ideas of how to start doing other things and finding other stuff on the Internet. Although now that you know how to find things for yourself, it won't take you long to find a whole lot more.

Grow a family tree!

How would you like to grow your very own family tree?

No, not that kind. The kind where you end up with a diagram that shows how all the people in your family are related to each other. You need to find out about family members from the past to do this. To start off with, you should really find out as much as you can by talking to your relatives – especially if your family is called Smith. You've got to narrow things down as much as you can. The most important thing is to know where in the world to start looking, and what names to look for. As with all the best research, it usually works well if you have a rough idea of what you expect to find. If you know your grandmother came from Hungary, then the record of a marriage between Daisy Featherstonehaugh and Roderick Fortescue-Smythe is unlikely to be part of your family history, even if your name is Smythe.

Be an Internet expert:
RESEARCHING ON THE INTERNET

WHAT TO DO

Try the www.familysearch.com site first. You may be lucky, and it's such an exciting site that it's always worth a visit.

If that's no help, there is a list of thousands of helpful family research sites all over the world on www.cyndislist.com

The UK and Ireland Genealogical Information Service (www.genuki.org.uk) has got lots of information and advice on how to do this kind of research. The UK Public Record Office also has information on line: www.pro.gov.uk.

Travelling

One of the most exciting things about the web has always been the fact that it's a *World Wide* Web. So where in the world would you like to go? You can use the Internet to book flights and hotels and so on – but I expect you're not really likely to go all that far if you've got to be home by tea-time. So why not try a bit of 'virtual' travel?

Be an Internet expert: TRAVEL THE WORLD

WHAT YOU'LL NEED

A map of the world – one with a fairly large scale if possible.

WHAT TO DO

Close your eyes, wave your finger above the map and, without peeping, point to a random place on the map.

Now use the Internet to find out all you can about that place, including what the weather is like, some history and how to get there. You may even be able to make contact with someone who lives there. Make sure you bookmark pages you like (add them to your 'Favourites') so that you can find them again.

A BIT OF ADVICE

Access to the Internet is still much more widespread in the richer countries of the world, so it may be tricky to find anyone to talk to in the middle of Africa, but you should be able to find out something about the world they live in. You might be able to find a pen friends' organization which has links there.

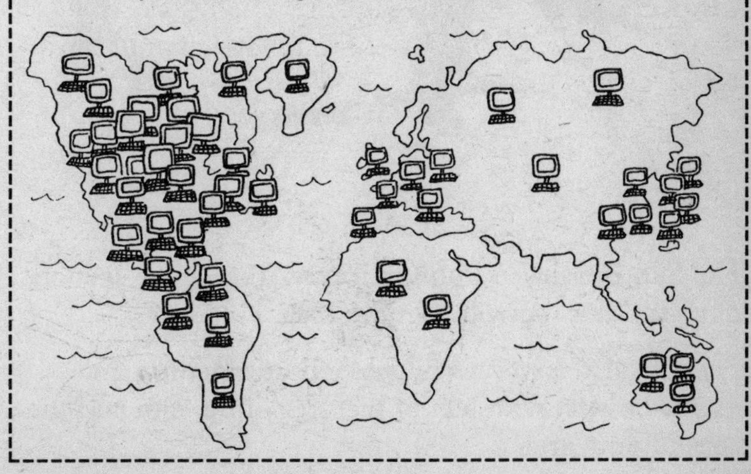

Try virtual travel ....

Here's a sample of stuff found in just 20 minutes of virtual travelling in San Francisco using Google (www.google.com) as the search engine.

Local newspapers like the *San Francisco Examiner* had lots of local and national news, traffic reports (there were roadworks on the Bay Bridge) and weather news – it was going to be a sunny day, with a number of rainy days to come.

San Francisco Zoo. Their white alligator is back and bigger than ever. A giraffe fell over and died the day before as its hooves were being trimmed. Nobody knows why.

The San Francisco Public Library. This has a section all about kids' activities in the area.

There is also a wonderful general information and what's-on site, with lots of pictures of the city and the surrounding area.

Museum sites. Try the Fine Arts Museum, or the City History Museum, which has some brilliant on-line displays about the 1849 Gold Rush, and the 1906 Earthquake.

There is loads of stuff about travel to and around the city, including hotels and restaurants, buses and tourist sites.

Webcams. These are cameras that send live pictures onto the web, including views of the sun coming up over San Francisco Bay.

On-line radio stations. There was a live early morning show (because the UK is eight hours ahead of California) on a really groovy jazz station.

Information Internet ....

Here is a list of a few of the amazing things you can find out about through the Internet.

You can find out all sorts of stuff about almost any place in the world – its history, what it looks like, what it's like to live there. Also how to get there. You can book flights, hotels, and car hire on the Internet. And you can ask people in chat rooms or newsgroups for tips on what to do when you're there.

In newspapers you can read all about different places, and find out what people there are thinking about. There are all sorts of newspapers and magazines on the Internet from all over the world, covering most things you might be interested in.

Some of them are electronic magazines. There are also lots of short stories and poems that are only published on the Internet.

You can even download whole books from the Internet. Project Gutenberg (www.gutenberg.net) is the most famous, but is only one of several projects to make (mostly out-of-copyright) books available in electronic form. You can download books by Dickens, Shakespeare and Mark Twain as well as thousands of other famous books. Because they're only text, the files aren't very big, so it won't take you too long to download them.

You can also visit museums on-line. Don't groan. Just think: you don't get sore feet, and you can listen to your music while you're finding out stuff, and looking at pictures. You can't get statues yet (apart from pictures of them) but I expect Internet holograms will come along soon. You can even visit 3-D reconstructions of places that you can move around in.

People have put webcams up all over the place so you can actually see what's going on – right this minute – in towns, in the countryside, even in people's living rooms. A good place to start looking for webcams is: www.earthcam.com.

CHAPTER SEVEN

ENTERTAINMENT

This is the bit you've all been waiting for. It's all very well using the Internet for serious stuff, but how do we get to all the cool exciting bits that we're all really interested in?

Sounds

It's absolutely amazing how many radio stations there are on the Internet. Some of them are ordinary stations which broadcast live on the Internet as well as over the air in their own country (Virgin Radio and Capital Radio in the UK, for example). Others are Internet-only stations – many of them run by people like you.

So you can now listen to the same radio station that a friend on the other side of the world listens to at breakfast every day. You can compare notes on the best music and the worst jokes. Magic. And because the Internet is multi-tasking (it can do more than one thing at a time), you can carry on doing something else on the computer while you're listening.

Be an Internet expert:
TUNE IN TO INTERNET RADIO

You need a couple of extra programs on your computer to get listening. Everything you need can be downloaded (transferred to your computer) free from the Internet.

WHAT YOU'LL NEED

 First of all go to www.real.com and download the Realplayer.

 Then go to www.microsoft.com/windows/mediaplayer and download the latest version of the Windows Media Player.

real player

media player

In both cases it's always worth having the most up-to-date version, because the improvements from one to the next can be quite amazing.

WHAT TO DO

Install your two players, and let them take you to their web sites – which they always do when you first install them. You'll find lots of links from both players to all sorts of radio stations, music and video sites. Both the Realplayer and the Windows Media Player have buttons on them that allow you to search for particular countries or kinds of music or radio stations.
If you haven't spent too much time on-line already, you can listen to a faraway radio station while you do something else on the Internet.

MP3

There's been a lot of hype about MP3 music on the Internet. In fact, it's apparently the most-searched-for term in search engines. So, what is MP3?

MP3 is short for 'MPEG layer 3'. You've probably heard of MPEG movies on DVD. MP3 is the audio part of video that has been MPEG compressed (digitally squeezed so it takes less space). To download a 5-minute piece of music from a normal CD could take over 6 hours. MP3 squeezes all that music, at nearly the same quality, into around a tenth of the size.

With MP3, 5 minutes of music only takes about 25 minutes to download.

The music industry is understandably worried because, as the Internet gets faster, download times will drop to something like real time (5 minutes of music downloaded in 5 minutes).

And why would anyone buy CDs if they could get the music on the Internet for free? Well, the answer is that record companies will eventually make official downloads easier, but charge for them.

The really brilliant thing about MP3 is that it means that if you've got a band of your own, you don't have to be signed to a record company to get your music out to your potential fans. You can just put it on the Internet. More on that sort of thing later.

In the meantime (finger-wagging, here) don't forget copyright. Stealing music (or anything else) from the Internet is just as much stealing as if you'd gone to a shop and stolen the CD from the shelf.

HOW TO PLAY MP3 FILES

There are dozens of free players out there for playing MP3 files. Both Windows Media Player and Realplayer will play them. There are other programs like Winamp, which convert sounds to MP3 and organize your collection for you.

If you're using these players for CDs as well, they can often look up the CDDB database for you. This is a database on the Internet that has details of millions of published CDs. The programs read the identity code on the CD in your computer, and find the title of the CD and all the tracks on it from the CDDB database. So next time you put the CD in, it'll tell you what it is and what song you're hearing.

Be an Internet expert:
PLAY AN MP3 FILE

You're going to find out how brilliant MP3 music can be.

WHAT TO DO
Make sure you've got one of the players installed. Go to MP3.com (www.mp3.com) and choose what kind of music you want to hear.

You can then play a low quality version while you're on-line to make sure you like it.

Now you can download the MP3 file to keep on your computer.

Movies

The computer in your bedroom is quite capable of originating CD-quality music. Back-bedroom musicians quickly found that the Internet was a perfect way of getting themselves heard.

The spread of fairly cheap Digital Video (DV) cameras, and the way in which the Internet is progressively becoming faster and faster, means that the same thing is beginning to happen to movie-makers. When digital-only cinema distribution is ready, movie-makers won't have to spend millions on shipping film copies around the world.

There are hundreds of independent movie-maker web sites out there already. Some of them are very good, some of them are complete rubbish, and most of them only show movies in a tiny little jerky box on the screen. But the experience is gradually getting better.

If you've been playing around with music and radio stations you might already have tried out some of the movie and TV sites. Realplayer and the Windows Media Player, together with Apple's Quicktime player (www.quicktime.com) are all you need.

There are two ways of dealing with moving pictures on the Internet. The way you're most likely to come across first is called 'streaming'. That means that the video (or audio) is sent bit by bit to your computer, just like a live TV broadcast.

Be an Internet expert:
EXPLAIN HOW 'STREAMING' WORKS

Try this experiment to explain how moving pictures are delivered to your computer.

WHAT YOU'LL NEED

 a funnel

 2 milk bottles

 water

WHAT TO DO

Fill one milk bottle with water, leave the other empty. Hold the funnel with your finger over the hole at the bottom while you fill the top with water from a tap.

Take your finger away, and quickly put the end of the funnel into the empty milk bottle. Begin pouring water from the full bottle straight away, before the funnel empties. You should be able to pour in such a way that the rate of water coming out of the bottom of the funnel will stay at a steady rate.

➤

EXPERT

As you pour the water into the funnel, you'll see that you can pour faster or slower, or even stop altogether for a while, but (for a second or so) the stream of water coming out of the bottom moves at a steady rate.

WHAT YOU PROVED

The top of the funnel is acting as what's called a 'buffer'. You might have noticed Realplayer saying it was 'buffering 5 seconds' or something similar. What it's doing is collecting a few seconds of movie before it begins playing, so that if the Internet connection slows down for a bit or gets interrupted, there's still some spare movie all ready to be played, and you don't notice any interruptions. It doesn't always work, of course, but it should keep movies running smoothly.

TO STREAM OR NOT TO STREAM?

Streaming is one way of watching movies on your computer. The other way of getting movies is to copy the whole movie to your computer before you watch it. Both ways have their advantages and disadvantages.

Streaming something obviously means that you see it straight away, without biting your fingernails down to the elbow with impatience.

On the other hand, you have to connect to the Internet to see it again.

Downloading a video or music file to your computer means that you can play it whenever you want. Many of the higher quality video clips are designed to be downloaded first, and not streamed, because you'd need a much faster connection to watch it in real time as you downloaded it.

Most of the time you don't have the choice. Either files can be streamed, or you can download them. Better quality is usually the reason you'd have to download something before watching it.

Be an Internet expert:
WATCH A VIDEO

WHAT YOU'LL NEED

 Make sure you've got one of the three movie players installed: Real Player, Windows Media Player, or Quicktime.

WHAT TO DO

Go to the home page of your movie player.
Look for some video, and click on it to watch it.
If you know the speed of your Internet connection, try clicking on something meant for faster connections and see what happens.

OTHER THINGS TO TRY

Two other sites which are very interesting are www.broadcast.com and www.pseudo.com. Once you start looking, it's amazing how much stuff there is.

Shockwave

For something else that moves, though not in quite the same way, try going to www.shockwave.com. Shockwave is a vector-animation format.

Don't panic. It's quite simple really. Pictures like your Windows background (assuming you're running Windows with a background) are drawn on the screen dot by dot. All the information about what colour each dot is supposed to be is what makes picture files big. The bigger the picture you're viewing, the bigger the picture file will need to be.

Vector graphics work differently. If you're drawing a vector picture of a box, all you have to do is describe where the corners are, and what colour it is. Since the file is just a set of instructions, it's much smaller than a file with an actual picture of a box would be. And you can make the box any size you like without the file getting bigger. It's a kind of magic.

So Shockwave is a way of producing animation (moving cartoon pictures) with very small file sizes. This makes it quick too, because all that is being sent over the Internet is a set of instructions, not a whole frame as it would be with video.

Shockwave sends out animated strip cartoons, like *South Park*, *Peanuts* and millions of others. It also sends out games you can play, and animated 'greetings cards' you can send to friends by e-mail. Try it out. A lot of the animated bits you see on all sorts of web sites all are made in this way.

Games

You can play all sorts of games on the Internet. Some of them you play against a computer somewhere – just like playing games on your own machine. But what's really exciting is that you can also play games against real people anywhere in the world.

There are thousands of game sites on the Internet. At many, you can play games like Chess or Draughts, or Noughts and Crosses, and even card games. You can also play different sorts of role-playing games, of the Dungeons and Dragons sort, where you are a dragon, a hero or heroine, or even an alien.

You can also shoot the blazes out of each other in on-line games – or whole tournaments – of games like Quake or the various Star Wars games.

HOW TO PLAY GAMES

In the past, people used to play games of chess by post (remember letters?). They didn't have to send the whole chess board and all the pieces through the post. Each player had their own board and all they had to send were details of the moves they made.

Playing games on the Internet works in a similar way. Everyone playing the game has their own copy of the game running on their own machine. Players only have to send information about their moves or what they're shooting, or whatever, through the Internet. This means the amount of data travelling backwards and forwards is kept to a minimum.

Normally there's some kind of chat facility, too, so that you can talk to each other – sometimes using your microphone, but more often using your keyboard – to arrange what rules you'll follow or where you'll start and that kind of thing.

Be an Internet expert: GAME ON!

You'll like this one.

WHAT TO DO
Go to the Microsoft Gaming Zone (http://zone.msn.com) and choose a game.

If you have the full version of one of the commercial games (like the Star Wars games) try joining a game and playing on-line against other people.

Next do some searches for other games sites. Look for: *game on-line*, for example, or *quake on-line, dungeons and dragons*, or *chess on-line*.

Shopping

The Internet is brilliant for certain kinds of shopping. The problem is that you need a credit card for most of it, and you have to be over 18 to get one of those. This is a problem for people selling things, too, so there's a lot of work going on to set up so-called 'micro-payments' systems.

With a micro-payment system, you would put some money into a special account (or persuade someone else to do it for you).

When you wanted to buy something you would give a special password for the money to be taken from this account. This will also be brilliant for sites that might want to charge you only 1p or 5p to download a piece of music, or subscribe to a newsletter or something like that. So far, it's been hard to collect tiny payments like this, but it will be a lot easier soon.

Internet shopping is best for stuff you can have straight away. A lot of computer software is now sold by download from the Internet. So is a lot of MP3 music – and soon movies on video will be available too. There's going to be a lot more sold this way, too, as soon as proper micro-payments systems get established. There are already a number of newspapers and magazines that have Internet sites. Well, as soon as they can easily charge for reading, say a penny a page, you can bet there will be a lot more.

SECURITY

A lot of people worry about giving credit card numbers over the Internet – and they are quite right to be careful.

But there are two things you can do to protect yourself.

1 Only buy from companies that you have some reason to trust, perhaps because they are well-known high street stores, or they are good brand names.

2 Make sure that the connection is secure. The page where you put the number should have 'https://' at the front of its name, and/or there should be a little key on the status bar at the bottom of your browser.

If you do these things then giving a credit card number over the Internet is really no more dangerous than giving it over the telephone, or to a waiter in a restaurant.

FREE STUFF

Never mind money. The Internet was built on the principle of sharing things around and not charging for them, and there's masses of free stuff you can download or get sent to you. Try an Internet search for 'freestuff' and see what you get.

There's lots of software out there. Some of it is free, and some of it is what's called 'shareware'. Shareware programs are free to try out, but if you like them and want to keep using them, you're supposed to pay a registration fee. Normally registering with it will give you a few more features and free up-dates. Good places to start for freeware and shareware are: www.tucows.net and www.zdnet.com.

CHAPTER EIGHT

DOING IT YOURSELF

OK, so now you know all about finding your way around, and you've been having a lot of fun. But how about putting your own site together, so that everyone can see what a massively talented person you are?

Think about it first

You'll have seen a few of them by now: those web sites which are all red writing on a purple background and the front page takes ages to download because it's all one enormous picture of someone's hamster (blurred), so you can't do anything at all until it's finished downloading. And are you interested by then? No, certainly not.

Everyone has their own preferences, but there are some things that good web sites do properly, in their different ways. And there are things that bad web sites do, too – like the purple hamster.

Things to think about

First of all, why should anyone want to visit your site? Why do you visit sites? To have a laugh, or find something you want, like information, music or pictures. When you set up your site, make sure you give people a reason to visit.

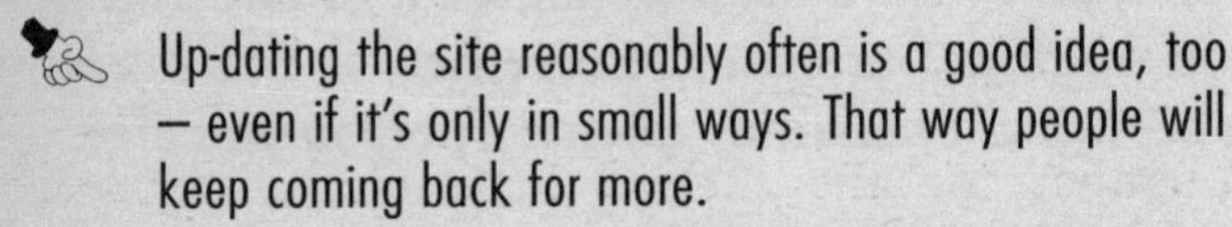

- Up-dating the site reasonably often is a good idea, too – even if it's only in small ways. That way people will keep coming back for more.
- Be thoughtful. Those wonderful graphics you took all weekend putting together might look terrific on your computer, but how long will people be prepared to wait for them to download? If they get bored they'll move off somewhere else. So make it fast. Or at least don't put them on the front page. That gives you a chance to tell people they might have to wait a while for some of the goodies, but that it'll be worth the wait.
- Finally, make it easy for people to find their way around. It's infuriating when you've found an interesting page, but then can't find your way back to it. Or worse, landing up on a web site you don't understand and can't find your way around.

Doing it ....

So, before you start putting a site together, you should decide:

- what is going to be on your site, and why
- how you are going to structure it so people can find things
- what it's going to look like

Remember, too, that the web likes things in bite-size chunks. It's better to have several small pages, logically divided up, than one horrendously long one.

SPEAKING WEB LANGUAGE

The language that makes web pages look the way they do is called 'html'. This is just text with instruction codes put in it. For example, if you want to use html to make something bold, you do this: <bold>**this sentence is bold**<bold>.

Step by step it's quite simple, but it can get quite complicated once you have a lot of it. An alternative way of putting together web pages is to use an 'editor', which looks more like a word processor. The editor can produce the code for you. This is probably a better way to start off.

If you have either Internet Explorer or Netscape, you already have a basic but quite useful web-page editor. Front Page Express or Netscape Composer are good enough to start with. There are also lots of others available on magazine cover disks and, of course, for downloading from the Internet.

Be an Internet expert: SET UP YOUR OWN WEB SITE

Imagine you and some mates have a band called Chuck the Munkie. You've been practising together in your garage for months now and your only audience has been the neighbour's cat and some rusty bike parts! Now, to give more potential fans the chance to learn to love your music, you're putting up a web site for it.

WHAT YOU NEED

The first thing you need is a plan of your site. What are you going to put on it? As you know, a web site is really just a number of pages that link to each other.

People usually call the first page 'index.htm'. You can call the others anything you like. Your front page could have some pictures and a menu linking to the other pages.

➤

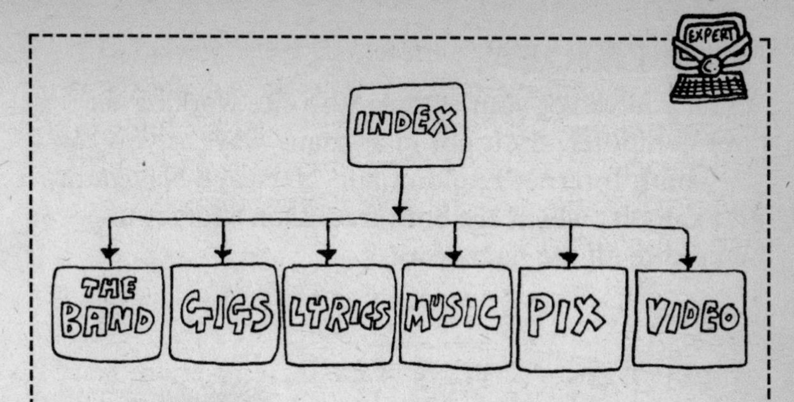

Here are some ideas for the other pages:

- a little piece about each of the band members
- details of where and when you're playing next
- sample lyrics from some of your songs
- pictures of you playing at gigs
- samples of some of your music
- maybe even a couple of videos: behind the scenes and playing a gig?

WHAT TO DO

Now put your site together. Your index page needs to include links to all the other pages, so people can find them. Make a link by highlighting the words you want people to click on ('lyrics', for example). Then find the 'make link' menu item in your web design program. It will ask you what page you're linking to. Tell it ('lyrics.htm', or whatever). Now the word 'lyrics' should be underlined on your index page, and when you click on it, the lyrics page will be loaded in your browser.

That's all there is to it. Now you can create a web site.

DOING IT FOR REAL

When you've got your sample web site working on your computer, try it out in as many ways as you can – try both Internet Explorer and Netscape Navigator, for example, which are both free. Don't forget to make sure all the links work.

When you're ready to show off to the world, you need to send your pages to your Internet Service Provider. Many web-page editing programs can do this for you, but you need to find out from your ISP where to send the pages. Usually there'll be a section of your ISP's web site called something like 'support', which will tell you what you need to do.

All programs are slightly different, so you'll have to spend some time exploring 'Help' menus to find out exactly how to do it.

Once you've got your site on the web, try it out again, and get all your friends to try it to make sure that it works on their machines.

Tips for making clear sites

The page title should be a (small) graphic, even if it's just words, so that it always looks the same, however people set their browser.

If you put several pictures on the page, make sure they're all small, so they load quite quickly. Visitors to the site will see that something is happening, which they might not if you have just used one big one.

Make sure your page is easy to read. This doesn't just mean getting the spelling right (*thow nobody is going to take you serriusly if you karnt even spel!*), but the colours, too. White on black looks good, but can be hard work to read. And as for red on black – well, you might just as well not have bothered.

Tips for tables

HTML is actually a pretty basic language. The only way you can be sure of laying things out exactly the way you want them is to use tables. Add a table from your web design program's menu, and put your pictures and words in different cells inside the table so they won't always spread across the screen and look messy.

Putting music on the web

A few pages of text will only need a tiny bit of the web space you get with your Internet account. Once you start putting music and video files on your web site, you need to think about how much space you really have.

Most Internet Service Providers give you around 20Mb of web space. That's only enough for about 20 minutes of MP3 music. You could make more of the space by putting some lower-quality streaming music on your web site. Both Real Networks and Microsoft give away free tools for turning the music you've recorded on your computer into a form that people can listen to across the Internet. (It's called *encoding*.) You can download a copy of Realproducer from Real Networks to help you do this.

If you go to www.mp3.com you'll find lots of software for encoding files to MP3 and various different players, too.

Putting video on the web

So, you fancy yourself as the next George (or Georgina) Lucas? Well, putting your videos on the web is as good a place to start as any.

Video is still a bit tiny and jerky on the Internet, but it's gradually catching up with music. Your video will look a lot better on the web if you use shots where there's not too much fast movement. You should also finish your editing with a movie with a picture size of something like 192x144mm.

Be an Internet expert:
MAKE A MOVIE!

WHAT TO DO

- Get Real Producer from www.real.com. It's a program that converts your movie for use on the Internet. Use it to 'encode' your movie.
- When it's finished, Real Producer will ask you if you want it to make a web page for your movie. Say 'yes', you can always edit it to make it look prettier later.
- Make a link to this web page from your index page. Now upload it all to your web space.
- And that's it. With enough disk space you can even produce your own TV channel!

Sticking with it ....

What everyone wants is 'sticky' sites: web sites that make people want to stay longer, and keep wanting to come back.

It's really not that hard to get all kinds of cool stuff to make your site sticky.

All you have to do is get to grips with the html code enough to insert a little bit from someone else's site into your page. If you can do that you can have a chat room, regularly up-dated news or a guest book. A guest book is a book you sign to leave a message saying you've been there and how brilliant you think the site is. Except on screen it looks more like e-mail than an actual book. You don't have to do anything too difficult yourself. All you're doing is linking to another web site, which does all the work for you, but makes it look like yours.

Another really good way of keeping people involved is to start a mailing list. You could even turn it into a regular e-mail magazine if you want. Listbot (www.listbot.com) will do all the work for you. You just put their button on your page.

If you go back to www.thefreesite.com, you'll find loads of this kind of stuff for webmasters (that's you). And with your well-practised search skills you won't have any trouble finding more.

Getting noticed

Having trouble getting noticed? Does it seem that no matter what you do, nobody drops by to visit your web site?

I'm afraid there's no magic wand you can wave to make people visit your site, but here are a few things you can do to help things along a bit:

- tell everyone you know, and get them to tell their friends, and so on
- make it such an interesting site that you can persuade your local newspapers to write about it
- go around the Internet to any similar sites you can find and get them to link to you in exchange for a link to them

- get yourself in the main web search engines: Altavista, Yahoo, Excite. There's a site called www.submit-it.com that will help you with others, and you can get programs to run on your own machine that will help you. The Exploit Submission Wizard is often on cover disks (or download it from www.exploit.net).
- get in touch with all the computer magazines and ask the people who write about web sites to look at yours

A few extras

So now you know it all. And if you don't know it all, then you certainly know how to find out – and that's all you really need to know.

Here are a few more site addresses to get you going. Have fun!

Start pages

Microsoft Network: www.msn.co.uk

Europe On-line: www.europeon-line.com

Yahoo: www.yahoo.com

Netscape: www.netscape.com

Search engines

Altavista: www.altavista.com

Excite: www.excite.com

Lycos: www.lycos.com

Google: www.google.com

Books and magazines

PC Zone: www.pczone.co.uk

BOL bookshop: www.bol.uk.com

Oxford University Press: www.oup.co.uk

Electronic Texts: www.etext.org

Internet Public Library: www.ipl.org

Movies and TV

Internet Movie Database: www.imdb/.co.uk

BBC: www.bbc.co.uk

The Farnham Film Company (author's site): www.farnfilm.com

Movies in the UK: www.popcorn.co.uk

Short films on-line: www.atomfilms.com

Disney: disney.go.com

Science

NASA: www.nasa.gov

Discovery Channel: www.discovery.com

Computing

Microsoft: www.microsoft.com

Apple: www.apple.com

Research Machines: www.rm.com

Games

www.gamesdomain.co.uk

www.gameplay.com

Travel

Good general travel site: www.a2btravel.com

Microsoft travel site: www.expedia.co.uk

Reference

Encyclopaedia Britannica: www.britannica.com

UK Phone Book: http://www.bt.com/phonenetuk

UK Postcodes & addresses: www.royalmail.co.uk/paf